YOUR KNOWLEDGE HAS VALUE

AF151494

- We will publish your bachelor's and master's thesis, essays and papers

- Your own eBook and book - sold worldwide in all relevant shops

- Earn money with each sale

Upload your text at www.GRIN.com and publish for free

Bibliographic information published by the German National Library:

The German National Library lists this publication in the National Bibliography; detailed bibliographic data are available on the Internet at http://dnb.dnb.de .

Imprint:

Copyright © 2015 GRIN Verlag, Open Publishing GmbH
Print and binding: Books on Demand GmbH, Norderstedt Germany
ISBN: 978-3-668-08580-0

This book at GRIN:

http://www.grin.com/en/e-book/309834/effect-of-cobalt-60-gamma-irradiation-on-storage-life-and-quality-of-kinnow

Saeed Ahmad, Waleed Iqbal, Muhammad Wadood Ehsan Ullah, Atif Nawaz

Effect of Cobalt-60 Gamma Irradiation on Storage Life and Quality of Kinnow Mandarins

GRIN Publishing

GRIN - Your knowledge has value

Since its foundation in 1998, GRIN has specialized in publishing academic texts by students, college teachers and other academics as e-book and printed book. The website www.grin.com is an ideal platform for presenting term papers, final papers, scientific essays, dissertations and specialist books.

Visit us on the internet:

http://www.grin.com/

http://www.facebook.com/grincom

http://www.twitter.com/grin_com

Effect of Cobalt-60 gamma Irradiation on storage life and Quality of Kinnow Mandarin

Dr. Saeed Ahmad, Waleed Iqbal*, Muhammad Wadood Ehsan Ullah and Atif Nawaz

Department of Horticulture, University of Agriculture Faisalabad

ABSTRACT

The current study was conducted to investigate the effects of Co-60 gamma irradiation on the storage life and quality of citrus (Kinnow mandarin). Fruits were irradiated with different doses of 0, 0.6, 0.8 and 1 KGy and stored at $4^{o}C$ with 90-95% relative humidity for 90 days. Fruit physical characteristics, chemical analysis and organoleptic evaluation were preformed before and after 30, 60 and 90 days of storage. The results of 'Kinnow mandarin' fruit treated with irradiation significantly showed better quality after storage as compare to controlled fruits. Among all the treatments the irradiation dose 0.6 kGy was found most appropriate dose to maintain the fruit quality during storage.

INTRODUCTION

Citrus is a non-climacteric fruit belongs to the class Dicotyledonae, order Sapindales, family Rutaceae and the sub-family is Aurantiodae. It consist of about 158 generas and 1900 species. Citrus is one of the most important fruit of the world which grown in more than 125 countries, It grows well at latitude 35°-36° with favourble climatic conditions under temperature range of $4^{o}C$-$50^{o}C$ (Mabberley, 2008). Citrus fruit is a rich source of vitamin-C, good source of amino acid, sugar, organic acid, and minerals (Niaz et al., 2004).

Citrus is the 2nd most important fruit of the world in terms of area and production after grapes. Pakistan at 11th position in world citrus production (FAO, 2009). In Pakistan, area under citrus cultivation is 194 thousand hectare and annual production is 2400 thousand tones (Anonymous, 2009). In Pakistan mostly cultivated citrus fruits are mandarins, oranges, grapefruit and lemons, and among these mandarins have larger share in production and export of total citrus fruits in Pakistan. Citrus has 33 percent share in total fruit production of Pakistan (Anonymous, 2010a). Pakistan exported more than 1346 tons of citrus annually to international markets (Anonymous, 2010b).

1

Due to the poor handling conditions the Kinnow fruits are more exposed to environmental stresses and more rapidly loses its quality. These conditions further intensify to the enhancement of physiological disorders such as, weight loss during marketing and storage, stem end rind breakdown, favor of mold attack and fruit rot etc. Internationally, several postharvest management paractices have been developed to control fruit disorders, maintain good quality, freshness and overcome the losses (Krochta, 1997; Hagenmaier, 2002; Bajwa and Anjum, 2007). One of these management paractices is the use of Gamma irradiation to enhance the shelf life and freshness of fruits and vegetables and reduce the risk of postharvest physiological diseases and disorders (Cia *et al.*, 2007). It is a process by which food is showed to a controlled source of ionizing radiation to maintain its shelf life and reduced food losses, enhance microbiologic safety, and reduce the risk of physiological disorders. This process is occasionally called as cold pasteurization because the microorganisms are inactivated at low temperatures (FAO, 1981).

Radiation doses are measured in international units called (Gray or Gy) and 1 Gy=100 radiation. The FDA has authorized the following 4 sources of ionizing radiation for food treatment: (Co-60), (Cs-137), machine-generated accelerated electrons not to exceed 10 MeV, and machine-generated X-rays not to exceed 5 MeV (Shea *et al.*, 2000). The use of irradiation with cold treatment is considered more effective (Beraha *et al.*, 1960; Tiryaki and Maden, 1991). Fresh fruits and vegetables are allowed to be irradiated at doses up to 0.1 KGy (Cia *et al.*, 2007). It can be used to overcome insect infestation in fresh vegetables and fruits, slow down postharvest ripening of fruits and reduce spoilage microbial activity from fresh vegetables and fruits (Brennand, 1995). Irradiation of food can overcome the threat of food borne illness (Rustom, 1997; Braghini *et al.*, 2009a). Irradiation decrease economic losses which are caused by food deterioration and enhances food safety, so this practice enhances approval of product exported by developing countries (Loaharanu, 1994).

The grapefruit shows very less injury when it is treated at 0.3 KGy (Miller and Mednold, 1996). The organoleptic properties, titratable acidity, Total soluble solids did not show any change even at dose treatment of 0.5 KGy. Citrus fruits are irradiated at dose level of (58-69) Gy to fruit fly disinfestations for quarantine purpose

2

(Hallman and Martinez, 2001). The oranges showed reduction in weight loss and increase the total soluble contents when its treated at 0.5 KGy (Khalil *et al.,* 2009).

The efficacy of irradiation on citrus is extensively accepted and mostly carried out with gamma radiation around all over the world. But, there is a huge gap in the knowledge and understanding of the effects of irradiation in the physical and organoleptic quality aspects of commercial citrus cultivar Kinnow mandarin of Pakistan.

It's a new knowledge and it is also critical because it is necessary to find out the optimum dose required to get safety and the less change in the quality of the fruit. This research paying attention of new advances targeted to the maintenance of physicochemical properties of Kinnow mandarin and as a result overcome the risk of postharvest decay. To date, there is a gap about the effect of irradiation treatment on an important foreign currency earn fruit of Pakistan. Thus, the purpose of this study was to identify irradiation level at which loss of Kinnow fruit quality is minimized and to evaluate the changes in the physical, sensory and chemical properties of Kinnow.

MATERIALS AND METHODS

The proposed study regarding the effects of Cobalt-60 gamma irradiation on storage life and quality of Kinnow mandrin (*Reticulata blanco* L.) was initiated on 22nd January 2014. The experiment was conducted at Postharvest department laboratory cold storage, Ayub Agriculture research Institute, Faisalabad. Uniform mature, fruits of Kinnow mandrin (*Reticulata blanco* L.) were harvested from Sq # 9, University of Agriculture, Faisalabad, Pakistan during 2014. After that fruits were packed in card board boxes and transported to Pakistan Radiation Services (PARRS) Foods Lahore for γ-irradiation treatment, absorbed dose was measured by using 'Harwell Amber 3042' dosimeters. In these studies, Cobalt-60 was used as the source of irradiation. The overall irradiation procedure involved keeping the palletized boxes of Kinnow in a chamber through a conveyer belt followed by exposure to the γ-irradiation from four sides of the pallets, for a defined period of time depending upon the dose of the treatment After completing the irradiation process, the boxes were removed from the irradiation chamber via the conveyer belt, and then after irradiation fruits stored at 4°C with 90-95% relative humidity in cold storage at post harvest lab Ayub Agriculture research institute Faisalabad.

Physical and biochemical analysis of fruits were performed in the Pomology Lab, Institute of Horticulture, University of Agriculture, Faisalabad. Each treatment was replicated three times Completely randomized design (CRD) under factorial arrangements (Treatments and storage days intervals) were used.

Treatments:

$T_1 =$ Control

$T_2 =$ Cobalt-60 gamma rays @ 0.6 KGy.

$T_3 =$ Cobalt-60 gamma rays @ 0.8 KGy

$T_4 =$ Cobalt-60 gamma rays @ 1 KGy.

Data collection:

A) Physical characteristics

1) Peel weight (%)

2) Peel thickness (mm)

B) Physiological characteristics

1) Fruit weight loss%

2) Decay (%)

C) Chemical characteristics

1) Total soluble solids (Brix)

2) TSS/ Acid ratio

D) Sugars

a) Reducing sugars

b) Non-reducing sugars

Statistical analysis

The experiment was carried out under Completely Randomized Design (CRD) along with factorial arrangements. The data recorded were analyzed using

RESULTS AND DISCUSSIONS

'Kinnow mandarin' fruit weight was considerably affected by different storage periods. Fruit weight reduced during storage and highest decrease in fruit weight was analyzed after 90 days which was due to loss of moisture through transpiration. During the whole storage period T_2 (0.6 kGy) dose

Analysis of variance technique with the help of computer run statistical program 8.1 and least significant (LSD) test was used to compare the treatments means (Steel *et al.*, 1997).

of radiation treated fruit maintain highest 'Kinnow mandarin' fruit peel weight.

It was detected that maximum average fruit peel thickness (3.27 mm) was observed in constant in (0.6 kGy) dose of irradiation as compare to other treatments. Fruits peel thickness keeps on decreasing with increased storage periods. The

4

minimum peel thickness of fruit observed after 90 days of storage (2.95 mm).

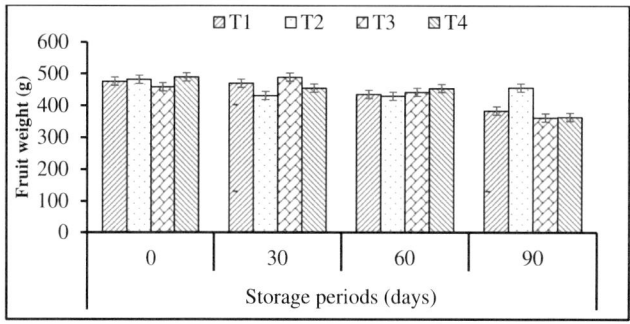

Figure 4.1: Effect of different postharvest treatments on fruit weight of 'Kinnow mandarin' during long term storage conditions. Values are means of three replications. Vertical bars indicate ± standard error (SE); where n = 3

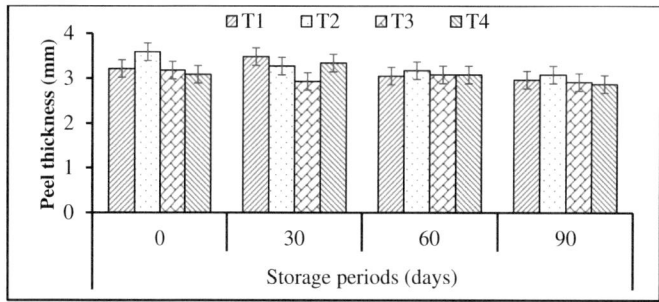

Figure 4.2: Effect of different postharvest treatments on peel thickness of 'Kinnow mandarin' fruit during long term storage conditions. Values are means of three replications. Vertical bars indicate ± standard error (SE); where n = 3

Higher physiological weight was occurred in T_1 (control) 'Kinnow mandarin' fruits with average physiological weight loss of (7.93%).Physiological weight loss percentage increased as storage period extended and maximum physiological weight loss being observed after 90 days of the storage period having mean physiological weight loss of (12.96%). It was perceived that highest fruit decay rate (17.77%) was detected in T_4 (1 kGy) irradiated 'Kinnow mandarin' fruit. while, lowest was noted in fruit treated with T_2 (0.6 k Gy) irradiated group (10.37%). Fruit decay rate was keeps on rising with enhance the storage interval. Fruit decay rate was continuously increased and the maximum fruit decay was observed after 90 days of cold storage intervals having mean fruit decay rate of (23.11%). Fruit decay was mostly due stem-end rot (*Botryodiplodia theobromae*) and Blue mold rot (*P. italicum*) during the whole storage period.

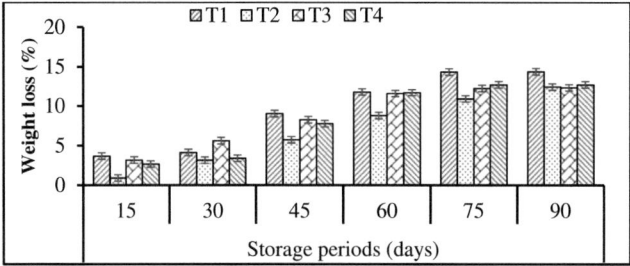

Figure 4.3: Effect of different postharvest treatments on physiological weight loss of 'Kinnow mandarin' during long term storage conditions. Values are means of three replications. Vertical bars indicate ± standard error (SE); where n = 3

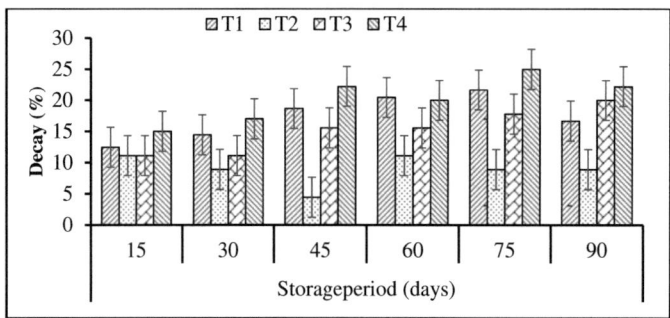

Figure 4.4: Effect of different postharvest treatments on fruit decay rate of 'Kinnow mandarin' during long term storage conditions. Values are means of three replications. Vertical bars indicate ± standard error (SE); where n = 3

Fruit under doses of radiation preserved higher total soluble solids as compare to control fruits and maximum total soluble solids were observed in fruit treated with (0.6 kGy) dose of radiation (9.08°Brix). Total soluble solids were decreased with the increase in the storage days and total soluble solids were reached at minimum value (8.36°Brix) after 90 days of storage. Higher total soluble solids in irradiated fruit due to breakdown of complex carbohydrates into simple sugars and thus it caused increase in the soluble sugars. Sugar acid ratio was increased with the increase in the storage periods in spite of and it was reached to maximum value after 90 days of storage (29.09). Interaction effects between treatments and storage periods not substantially affected sugar acid ratio of 'Kinnow mandarin' fruit juice.

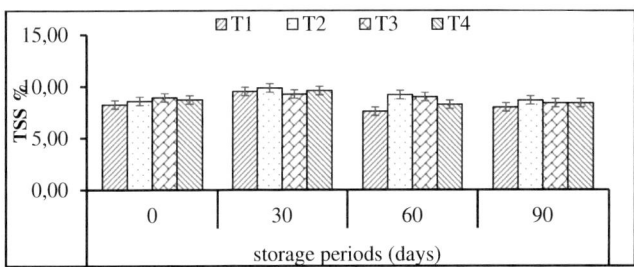

Figure 4.5: Effect of different postharvest treatments on TSS of 'Kinnow mandarin' fruit during long term storage conditions. Values are means of three replications. Vertical bars indicate ± standard error (SE); where n = 3

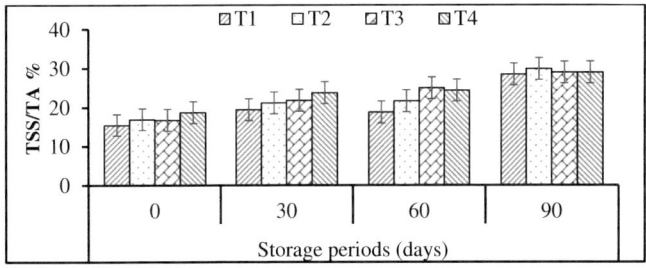

Figure 4.6: Effect of different storage periods on TSS/Acid ratio of 'Kinnow mandarin' fruit during long term storage conditions. Values are means of three replications. Vertical bars indicate ± standard error (SE); where n = 3

All the treated fruits have more reducing sugars as compare to control fruits, Reducing sugar was increased from (0.6 to 1) kGy and highest reducing sugars were recorded in fruit treated with (1 kGy) dose of irradiation. Reducing sugar was increased with the expansion of storage durations and attain the maximum amount of reducing sugar contents after 90 days storage (12.51%). Non-reducing sugars of Kinnow mandarin fruit juice showed significantly differed results for all the treatments and highest non-reducing sugars were observed in T_2 (0.6 kGy) fruits (5.80%). Non-reducing sugar was gradually decreased with increase in the storage period and non-reducing sugars reached at lowest value (2.40%) after 90 days of storage.

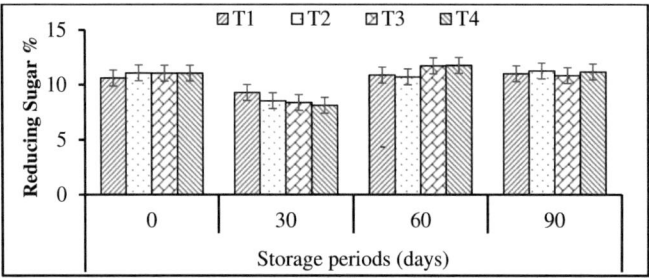

Figure 4.7: Effect of different postharvest treatments on reducing sugars of 'Kinnow mandarin' fruit during long term storage conditions. Values are means of three replications. Vertical bars indicate ± standard error (SE); where n = 3

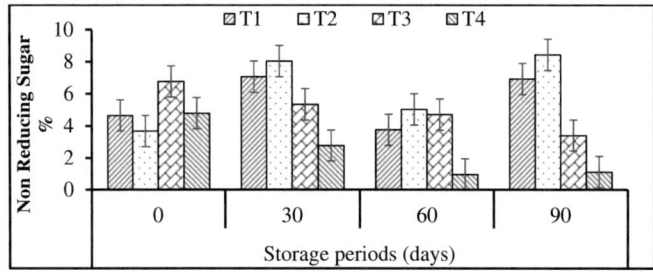

Figure 4.8: Effect of different postharvest treatments on non-reducing sugars of 'Kinnow mandarin' fruit during long term storage conditions. Values are means of three replications. Vertical bars indicate ± standard error (SE); where n = 3

References

Bajwa, B.E. and F.M. Anjum, 2007. Improving storage performance of *Citrus reticulata Blanco* mandarins by controlling some physiological disorders. Int. J. Food Sci. Tech, 42:459-501.

Beraha, L., G.B. Ramsey, M.A. Smith and W.R. Wright. 1960. Gamma radiation dose response some decay pathogens. Phytopathol. 50:474-476.

Cia, P., S.F. Pascholati, E.A. Benato, E.C. Camili and C.A. Santos. 2007. Effects of gamma and UV-C irradiation on the postharvest control of papaya anthracnose. Postharvest Biol. Technol. 43:366-373.

FAO.1981. Wholesomeness of irradiated food. Technical Report Series 659.

FAO. 2009. Food and Agriculture Organization of United Nation. (Available online with updates at http://faostat.fao.org/default.aspx? Page ID=567).

Hagenmaier, R.D. 2002. The flavor of mandarin hybrids with different coatings. Postharvest Biol. Technol. 24:79-87.

Hallman, G.J. and L.R. Martinez. 2001. Ionizing radiation quarantine treatment again Mexican fruit fly *(Diptera tephritidae)* in citrus fruits. Postharvest Biol. Technol. 23:71-77.

Khalil, S.A., S. Hussain and M. Khan. 2009. Effect of gamma irradiation on quality of Pakistani blood red oranges (*Citrus sinensis* L.). Int. J. Food Sci. Technol. 44:927-931.

Krochta, J.M., 1997.Edible composite moisture barrier films Packaging Year Book. In: B. Blakiston,(Ed). 1996.Washington DC.

Mabberley, D.J. 2008. A Portable Dictionary of Plants: their Classification and Uses. 3rd Ed. Cambridge University Press.

Miller, W.R. and R.E. Medonald. 1996. Post harvest quality of GA- Florida grapefruit after gamma irradiation with TBZ and storage. Postharvest Biol. Technol. 7:53-260.

Niaz, A.C., M.N. Maken and S.A. Malik. 2004. Native home historical background and importance of citrus fruits in Pakistan. p. 48-56. In: M. I. Chaudhary and R. Anwar (eds.). Proc.1st Int. Conf. on Citriculture 28-29 April. University of Agriculture, Faisalabad, Pakistan.

Shea, K.M., E.A. Benato and E.C. Camili technical Report: Irradiation of Food. American Academy of Pediatrics, Pediatrics.106:1505-1510.

Steel, R.G.D., J.H. Torrie and D.A. Dicky. 1997. Principles and Procedures of Statistics. A Biometrical Approach, 3rd ed. McGraw Hill Book Co. New York.

Tiryaki, O. and S. Maden. 1991. Penicillium expansum, *Botrytis cinerea* ve *Rhizopus nigricans* Effect of Gamma radiation on Ankara pear fruit. Food Chem. 6:7-11.